美国国家地理 儿童小百科

NATIONAL GEOGRAPHIC KiDS

雪豹和狮子
Snow Leopards & Lions

[美] 吉尔·伊斯巴姆 [美] 艾米·斯凯·科斯特 / 著 张佩芷 / 译

四川少年儿童出版社

雪豹
Snow Leopards

看，一只雪豹！
A snow leopard!

　　它在布满岩石的山脊上漫步。在这里，嫩芽意想不到地从裂缝里冒出来。

　　它穿过阳光斑驳的沟壑，跃过缓缓流淌的融雪的溪流，窥探着洒满露水、生机盎然的草地。

She wanders rocky ridges, where sprouts push from unlikely cracks.

She crosses sun-dappled ravines, leaps at trickles of melting snow, and explores dewy meadows buzzing with life.

它在找寻着什么?
　　它在找一个洞穴。它找到一个藏身之处，在这里可以躲避寒风冰雪。

What's she looking for?
A den. She finds a hidden place, protected from icy winds and swirling snow.

在这里,它将会产下两三只小雪豹。

Here, she'll give birth to two or three cubs.

依偎在一起
Nuzzle

小雪豹紧靠着雪豹妈妈,感到十分安心。刚出生的小雪豹眼睛都还没睁开,十分虚弱,还不能四处活动。

不过,一周左右……

早上好,这个世界!

Mom stays close so the cubs know they're safe. Newborns are born with their eyes closed, too weak to move around much.

But, a week or so later . . .
Good morning, world!

特别的猫科动物
Special Cats

孟加拉虎（生活在亚洲）
Bengal tiger Asia

大多数猫科动物的眼睛是黄色或者金色的。雪豹的眼睛却是灰色或者淡绿色的。

Most big cats have yellow or gold eyes. A snow leopard's eyes are gray or pale green.

你的眼睛是什么颜色的?
What color are your eyes?

豹（生活在非洲、亚洲）
leopard Africa, Asia

非洲狮（生活在非洲）
African lion Africa

你会咆哮吗?
Are you quiet or LOUD?

你有胡须吗?
Do you have whiskers?

大多数大型猫科动物都会咆哮，但是雪豹不会。它们会发出低吼声、嘶嘶声、喵喵声、呻吟声、呼噜声、号叫声和喘气声。

Most big cats roar. Snow leopards cannot. Instead, they growl, hiss, mew, moan, purr, and yowl. They also make a puffing sound called a chuff.

你能像雪豹一样发出嘶嘶声吗?
Can you hiss like a snow leopard?

美洲狮
（生活在北美洲和南美洲）
mountain lion
North and South America

扑过去!
Pounce!

小雪豹在洞穴里长到两个月大时,已经可以玩耍了。它们翻滚着扭打成一团。

After two months in the den, cubs are ready to play! They roll and tumble and bat at each other.

直到夏末,小雪豹都跟着雪豹妈妈,寸步不离,去熟悉岩洞周边的各个角落。

By late summer, the cubs are following their mom everywhere, learning every nook and cranny of their rocky neighborhood.

雪豹喜欢立在高处，这样一来，它们就可以望得很远。啊……太阳下，岩石凸出地面的部分就是它们最好的歇息处。

Snow leopards like to be up high, where they can see a long way. *Ahhh* . . . a rocky outcropping is a great place to perch in the sun.

当雪豹妈妈发现胖乎乎的旱獭时，它别提有多开心了。它一路追着旱獭穿过高高的草丛和流动的溪流，又冲下雪坡——它在给宝宝们寻找食物呢。

But if Mom sees a plump marmot, she's more than happy to chase it through tall grasses! Along a stream! Down a snowy slope! Her growing cubs need to eat.

午饭吃什么?
What's for Lunch?

你吃肉吗?
Do you eat meat?

你能捉住小鸟吗?
Could you catch a bird?

雪豹是肉食动物,它们会靠一只野羊或山羊来度过一周。

Snow leopards are meat-eaters. They will feast on a wild sheep or goat for a week.

当雪豹捕猎的时候,它们会发出声音还是保持安静呢?
While hunting, is a snow leopard quiet or noisy?

你喜欢吃绿色蔬菜吗?
Do you like green food?

喜马拉雅蓝羊（生活在印度）
Himalayan blue sheep India

你有过在雪地里行走的体验吗？
Have you ever walked through snow?

在雪豹的食谱上还有一些小型动物，比如旱獭、鼠兔、野兔和鸟类。雪豹还会吃草和树枝呢！

Also on the menu are smaller animals like marmots, pikas, hares, and birds. Snow leopards eat grass and twigs, too!

鼠兔（生活在中国）
pika China

金旱獭（生活在印度）
golden marmot India

寒冷的秋夜将山上的草丛变得干枯易碎。野羊和山羊们选择下山过冬。

Cold autumn nights turn mountain grasses dry and brittle. Wild sheep and goats move down the mountain for the winter.

亚洲巨角塔尔羊（生活在中亚）
Asiatic ibex Central Asia

雪豹悄悄跟着它们。

　　明年，小雪豹们要开始捕猎了。它们正在观摩和学习妈妈如何捕获猎物。

Snow leopards follow.

Next year, the cubs will begin hunting. For now, as they watch Mom sneak after her prey, they are learning.

白天狂风大作，夜晚天寒地冻，山里的冬天总是那么难熬。

Blustery days? Freezing nights? Mountain winters are harsh.

雪豹必须寻找可以躲避寒风和呼啸的暴风雪的地方。

Snow leopards must seek shelter from icy winds and howling blizzards.

寒夜茫茫!
Frosty Nights!

在寒冷的夜晚，什么东西能让你温暖？
What warms you on a cold night?

雪豹身上特别厚实的皮毛能保暖。它将长长的尾巴覆盖在柔嫩的鼻子上，沉沉睡去。
晚安，雪豹！

Extra thick fur keeps snow leopards warm. Long tails curl around to cover tender noses.
Nighty-night!

你的耳朵边缘是尖的还是圆的？
Are the tops of your ears round or pointy?

圆圆的小耳朵利于保持体温。
Small, rounded ears help hold in body heat.

你的鼻子是热的还是冷的?
Is your nose warm or cool?

毛茸茸的爪子让脚趾暖和舒适。
Furry paws keep toes toasty.

怒吼!
Growl!

兀鹫和渡鸦试图偷走一只雪豹的食物。在这危急时刻，这些看起来毛茸茸的"大猫"就变成了凶狠的斗士……

Vultures and ravens trying to steal a snow leopard's food soon find out that these cuddly-looking cats can be fierce fighters...

特别是当雪豹妈妈觉得它的孩子们有危险的时候,它会提高警惕!

. . . especially if a mom thinks her cubs are in danger. Then watch out!

25

小雪豹两岁前都在妈妈身边生活。在这些小雪豹长大和独立之前,雪豹妈妈不会再生孩子。

Cubs stay with their mom for nearly two years. She won't have more cubs until these are grown and gone.

当这些小雪豹长大后，它们会离开妈妈。它们必须找到属于自己的领地。

When the young snow leopards are finally grown, they leave. Each must find a new neighborhood, or territory, to call its own.

跳跃吧，雪豹!
Leap, snow leopard!

小雪豹在属于它的新的狩猎场上自由奔跑，茁壮成长。

Live strong and free in your new hunting grounds.

雪豹的栖息地
Where Snow Leopards Live

雪豹生活在亚洲大陆的高山地带。

Snow leopards live in the mountains on the continent of Asia.

地图指南
MAP KEY

橙色表示雪豹的栖息地
Where snow leopards live

指出雪豹的身体部位
Name the Parts of a Snow Leopard

你能指出图中雪豹的各个身体部位吗?

Can you find all the parts of a snow leopard?

爪子	眼睛	耳朵
• paws	• eyes	• ears
尾巴	鼻子	肚子
• tail	• nose	• tummy

狮子
Lions

看，一只狮子！
A lion!

它迈着带肉垫的大爪子，穿过长满草的平原。它一边漫步，一边四下张望、凝神静听。

He strides across the grassy plains on big, padded paws. Looking and listening, he keeps watch as he roams.

它漫步走过岩石，在水坑边躺下，蜷缩起来等待这场骤雨停歇。

He wanders over rocks and lounges by a water hole. He huddles low to wait out a sudden rain.

狮子的头和脖子周围的长毛叫作鬃毛。只有雄狮才会长鬃毛。

The long fur around a lion's head and neck is called a mane. Only male lions grow manes.

咆哮！
Roar!

这只狮子的职责是保护它的家族和领地。

This lion's job is to protect his family and their space.

当其他狮子靠近，它就会大声咆哮，警告入侵者"严禁入内"！

When lions from other families get too close, he roars loudly. It warns "Keep out!"

它的咆哮声吵醒了正在打盹儿的家族成员们。它们一个接一个地打哈欠、伸懒腰。当它也跟着伸懒腰时，它们却发出了低沉的呼噜声。

The lion's roar also wakes his family from a nap. They yawn and stretch. They grunt and growl as he joins them.

狮子家族被称为"狮群"。一个狮群一般有两到三只雄狮，几只雌狮和它们的幼狮。

Lion families are called prides. A pride usually has two or three males and several females and their cubs.

狮子是唯一的群居野生猫科动物。

Lions are the only wild cats that live in family groups.

到雌狮们为狮群准备晚餐的时间了,母亲们、女儿们、姐妹们组队去猎杀大型动物。

It's time for the females to gather dinner for the pride. Mothers, daughters, and sisters team up to hunt big animals.

从黄昏到黎明，它们悉心寻找猎物。

水坑附近是捕猎的最佳地点，可要当心鳄鱼哟！

They roam from dusk to dawn, on the lookout for prey.

Water holes are great places to catch food. But watch out for crocodiles!

角马
wildebeest

猛扑!
Pounce!

突然，一只狮子从藏身处冲出，向它的猎物猛扑过去。

今天能吃到什么呢？是角马，斑马，还是羚羊？

A lion sneaks up on her prey. She dashes at it from her hiding place.

What will today's meal be?
Wildebeest?
Zebra?
Gazelle?

羚羊
gazelle

雌狮既是凶猛的猎手，也是温柔的母亲。

Female lions are fierce hunters, but they are tender mothers.

狮子妈妈将刚出生的小狮子们藏在灌木丛里。它用嘴轻轻地叼起其中一只小狮子。

A mother lion hides her newborn cubs in the bushes. To move one, she carries it gently in her mouth.

刚出生的小狮子有斑点状的皮毛。随着小狮子逐渐长大，这些斑点就会逐渐消失。

Newborn lion cubs have spotted fur. The spots fade away as the cubs grow.

狮子妈妈用舌头将小狮子们舔干净。

它给小狮子们喂奶。

A mother lion licks her cubs clean with her tongue.

She feeds them with her milk.

它教小狮子们如何在野外生存。

She shows her cubs everything they need to know to survive in the wild.

玩耍时间!
Playtime!

小狮子们跟你一样，也爱玩耍。它们喜欢玩树枝和——尾巴！

Just like you, growing cubs love to play. They play with sticks—and tails!

它们会爬到妈妈或者狮群中其他成年狮子的背上玩耍。

They clamber over their mothers and other adults in the pride.

小狮子们互相追赶、跳跃、扑腾、撕咬，在草地上扭打成一团。

Cubs chase, leap, pounce, and bite. They tumble and tussle in the grass.

对小狮子们来说，玩耍不仅仅是为了开心，同时也教它们学会寻找猎物和保护自己的本领。

Playing is more than just fun for little lions. It also teaches them how to hunt and how to defend themselves.

呼噜呼噜!
Snore!

　　闲逛、咆哮、打猎、打闹和玩耍之后，狮子们打盹儿的时间又到了。草地上，狮群靠在一起睡觉。好梦，狮子们！

After all of that roaming, roaring, hunting, pouncing, and playing, it's time for another nap. The pride snuggles together in the grass. Sleep well, lions!

天生猎手
Built to Hunt

皮毛：皮毛的颜色方便狮子隐藏在高高的草丛中，等待猎物靠近。
FUR The color of its fur helps a lion blend in with tall grass as it waits for its food to come closer.

尾巴：长长的尾巴帮助狮子在奔跑和跳跃的时候保持平衡。
TAIL Its long tail helps a lion stay balanced as it runs and leaps.

后腿：修长的后腿很适合跳跃。
BACK LEGS The lion's long back legs help it jump.

在狮群中，雌狮承担着大部分的捕猎工作。它们猎取角马、斑马等大型动物，以及疣猪、羚羊等较小的动物。图中这些身体特征能帮助狮子追赶和捕获猎物。

Female lions do most of the hunting for a pride. They hunt large animals such as wildebeests and zebras, and smaller animals such as warthogs and gazelles. Here are a few things that help lions chase and catch the food they eat.

狮子是唯一在尾巴末端有一撮毛的大型猫科动物。
The lion is the only big cat with a tuft of fur at the end of its tail.

你能说出几种其他的野生猫科动物吗？
How many other wild cats can you name?

眼睛：狮子的眼睛可以看到远处的动静，即使在夜晚也视力绝佳。
EYES Its eyes can see movement far away. Lions can also see well at night.

下颌：强有力的下颌和又长又尖的牙齿帮助狮子抓咬猎物。
JAWS Powerful jaws and long, sharp teeth help lions grab hold of food.

脚：带肉垫的脚掌让狮子可以一声不响地靠近猎物。
FEET Padded feet help a lion creep up quietly on its food.

你能不出声地行走吗？
How quietly can you walk?

爪子：锋利的爪子可以紧紧地抓住猎物。
CLAWS Sharp claws snag and grip prey.

我们的叫声也很大
Roars and More

和其他狮子交流的时候，狮子会发出低吼声、嘶鸣声和呼噜声，不过最厉害的还是它们震耳欲聋的咆哮声。狮子的咆哮声可以传到5英里（8千米）之外。这里还有一些动物，它们有的体形很小，但能发出巨大的声音。

Lions growl, snarl, hiss, and grunt to talk to each other, but they are famous for their earsplitting roars. A lion's roar can be heard up to five miles (8 km) away. Here are a few more animals—some of them quite tiny—that can really raise a racket.

油鸱 是最吵闹的鸟，它们住在洞穴里，总是咯咯尖叫。
Oilbirds are the loudest birds. They live and shriek and squawk in caves.

小小的**灌丛蟋蟀**发出的唧唧声跟电锯一样吵。
The tiny **bushcricket's** chirp can be as loud as a power saw.

吼猴的叫声在3英里（5千米）之外都能听见。
A **howler monkey's** call can be heard up to three miles (5 km) away.

蓝鲸是声音最大的哺乳动物，它在水下发出的声音可以传500多英里（805千米）远呢。
The **blue whale** is the loudest mammal of all. Its underwater song can be heard more than 500 miles (805 km) away.

当**考齐蛙**一齐鸣叫，吵闹程度堪比一台除草机！
When **coqui frogs** call together, they can be as noisy as a lawn mower.

你能说出其他很吵的动物吗？
Can you name some other loud animals?

你能发出多大的声音？
How loud a noise can you make?

你能说出一些安静的动物吗？
Can you name some quiet animals?

59

狮子的家园
Home of the Lion

狮子生活在非洲和亚洲的部分地区。
Lions live in parts of Africa and Asia.

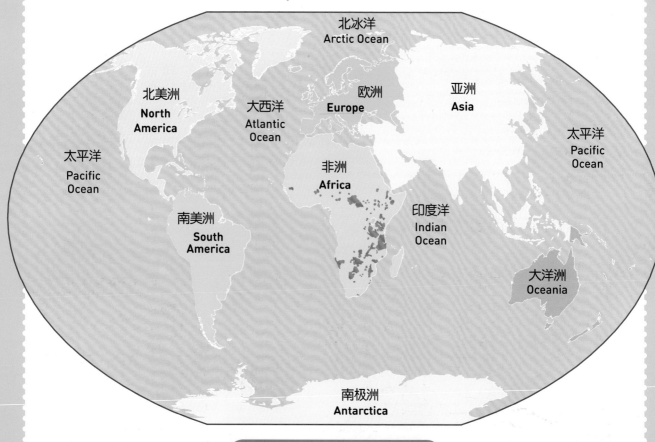

地图指南 MAP KEY

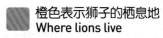 橙色表示狮子的栖息地
Where lions live